BOOKWORMS

Animal Structures

Mouths, Teeth, and Tongues

Derek Miller

Cavendish Square
New York

Published in 2019 by Cavendish Square Publishing, LLC
243 5th Avenue, Suite 136, New York, NY 10016

First Edition

Library of Congress Cataloging-in-Publication Data

Names: Miller, Derek L., author.
Title: Mouths, teeth, and tongues / Derek Miller.
Description: First edition. | New York : Cavendish Square, [2018] | Series: Animal structures | Audience: K to grade 3.
Identifiers: LCCN 2018013850 (print) | LCCN 2018017436 (ebook) | ISBN 9781502642356 (ebook) | ISBN 9781502642349 (library bound) | ISBN 9781502642325 (pbk.) | ISBN 9781502642332 (6 pack)
Subjects: LCSH: Mouth--Juvenile literature. | Animals--Adaptations--Juvenile literature. | Anatomy--Juvenile literature. | Adaptation (Biology)--Juvenile literature.
Classification: LCC QL857 (ebook) | LCC QL857 .M567 2018 (print) | DDC 591.4/4--dc23
LC record available at https://lccn.loc.gov/2018013850

Editorial Director: David McNamara
Copy Editor: Nathan Heidelberger
Associate Art Director: Alan Sliwinski
Designer: Megan Metté
Production Coordinator: Karol Szymczuk
Photo Research: J8 Media

The photographs in this book are used by permission and through the courtesy of: Cover (left to right) KarSol/Shutterstock.com, Martin Prochazkacz/Shutterstock.com; Hello RF Zcool/Shutterstock.com; p. 5 Doroty/Shutterstock.com; p. 7 Linda Marie Caldwell/iStockphoto.com; p. 9 Zoonar GmbH/Alamy Stock Photo; p. 11 Vmargineanu/Shutterstock.com; p. 13 Alex Ugalek/iStockphoto.com; p. 15 Maly Designer/Shutterstock.com; p. 17 Buddy Mays/Corbis/Getty Images; p. 19 Paul Hakimata Photography/Shutterstock.com; p. 21 Michel & Christine Denis-Huot/Science Source.

Printed in the United States of America

Contents

Animals need food and water.

A **mouth** lets an animal eat.

A mouth lets an animal drink.

Teeth help animals eat.

Teeth break up food.

This makes eating easy.

Flat teeth chew plants.

Sharp teeth rip meat.

People can eat plants and meat.

They have flat teeth and sharp teeth.

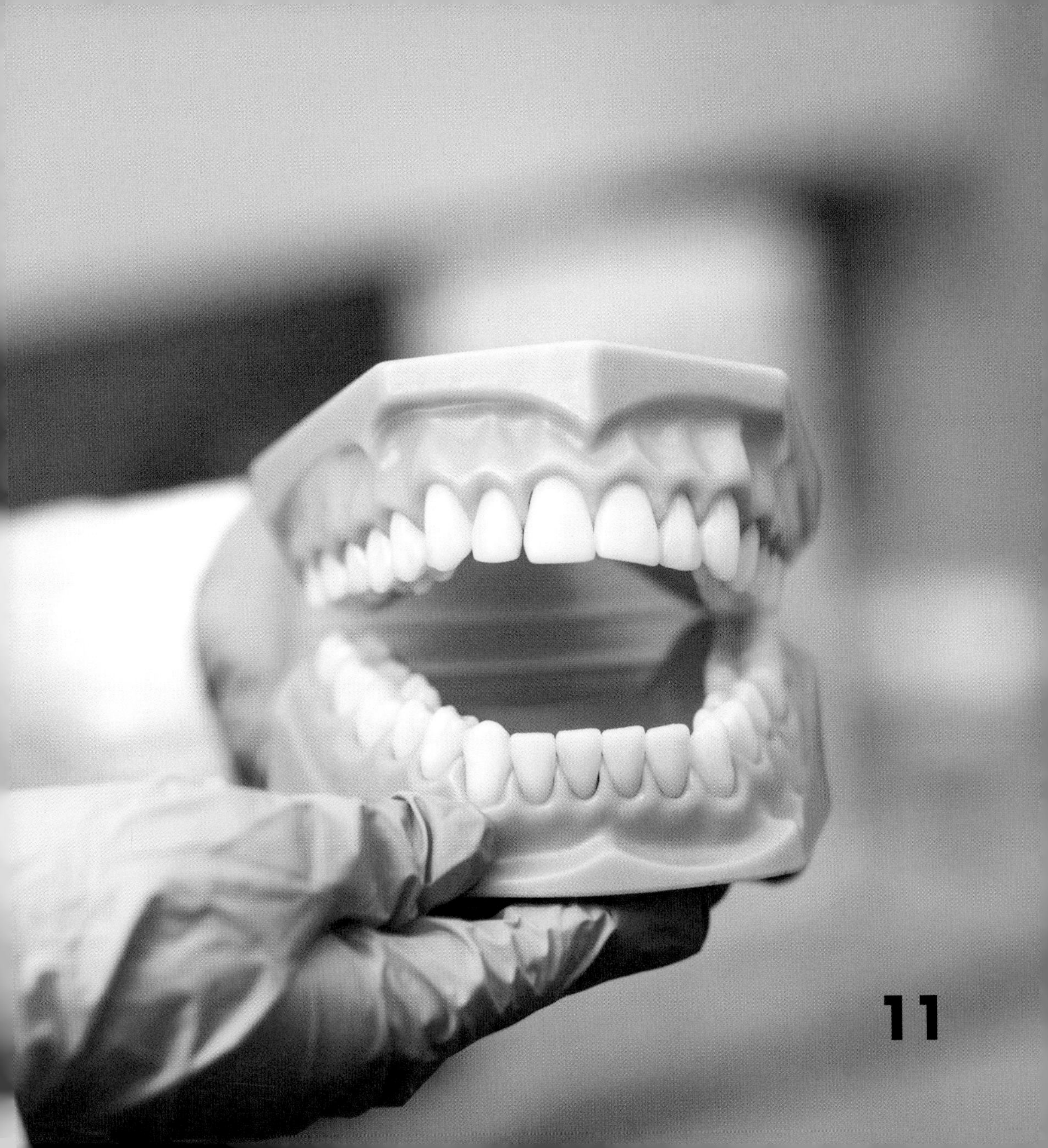

A **tongue** helps an animal eat.

Tongues move food around.

They help drink.

They lick up water.

Tongues can clean.

Cats and dogs lick their **fur**.

They lick away dirt.

Long tongues can catch food.

Frogs grab bugs.

Mouths can make sounds.

Dogs bark.

Snakes hiss.

People sing.

Mouths, teeth, and tongues
are important.

They let animals eat,
drink, and speak!

New Words

fur (FUHR) The hair of an animal.

mouth (MOWTH) The opening where food goes into the body.

teeth (TEETH) The sharp or flat parts in a mouth. Teeth are used to chew.

tongue (TUHNG) The soft muscle in a mouth.

Index

About the Author

Derek Miller is a teacher and writer. He is the author of many books for young students. He likes to write about animals.

About BOOKWORMS

Bookworms help independent readers gain reading confidence through high-frequency words, simple sentences, and strong picture/text support. Each book explores a concept that helps children relate what they read to the world they live in.